ZERO'S NEXT-DOOR NEIGHBOR

IMAGINING THE EXISTENCE
OF THE SMALLEST NUMBER,
THE LARGEST NUMBER,
AND THE SUPERLATIVE NUMBER SYSTEM

THOMAS J. PRESTOPNIK

Visit Thomas J. Prestopnik's website at
www.TomPresto.com.

Cover Art & Illustrations: Thomas J. Prestopnik

Zero's Next-Door Neighbor
ISBN-13: 978-1722282370
ISBN-10: 1722282371

CreateSpace Independent Publishing Platform
North Charleston, SC

Printed in the United States of America

From a small seed a mighty trunk may grow.

Aeschylus

NUMBER SYSTEMS

(and some elements of each)

Natural (1, 2, 3)

Whole (0, 1, 2, 3)

Integers (-2, -1, 0, 1, 2)

Rational (-⅓, ½, ⅗, ¾)

Irrational ($\sqrt{2}$, e, π)

Real (-2, -¾, 0, 1, $\sqrt{2}$, ½, π)

Imaginary (-3i, i, $i\sqrt{5}$)

Complex (½ - 3i, 4 + 10i)

Superlative (-K, -5α, 0, α, ½, 8 + 14α, 27, K)

CONTENTS

INTRODUCTION

The Sweet Scent of Pi

π

This is a small book about a very small idea. (And a very large idea, too.) Namely, the existence of the smallest number.

Obviously, a smallest number cannot exist in decimal form. For any given chain of zeros to the right of a decimal point, which is then followed by the number one–such as, 0.00000000000000000000000000001–you can always divide that number by 2 (or 3 or 4 or 5) to get an even smaller number. Add a million, billion or trillion zeros after a decimal point (or even an octobazillion zeros, should such a number exist) followed by a one, and the same rule applies.

So, the smallest number I'm referring to is an abstract concept, similar in a way that i is equal to $\sqrt{-1}$. As there is no real number to represent the square root of -1, the imaginary number i was invented and expanded upon. I have taken the notion of a smallest number along a similar path with a few (okay, maybe a *lot* of) hurdles remaining.

Which brings me to my first of two reasons for writing this book. One is simply to present the idea of a smallest number to any who might find it of interest, an idea that has been percolating inside my mind for a little over thirty years. (And neglected for months and years at a time over that same period, too.) But because of my limited math skills, I've only been able to develop this concept to a small degree. I've planted the seed of an idea and even got it to sprout a tiny bit, but that's about as far as my algebraic farming know-how can take it.

So, secondly, to grow this mathematical crop to whatever extent possible, it needs the expertise of some math brainiacs, maniacs, wizards and warriors to peek under the hood (pardon the mixing of metaphors) and tinker with the invented numbers to see if this numerical engine can get started.

This idea is here for others to examine (or to mock, totally your choice) who may find the topic of interest and might wish to expand upon it as a theoretical exercise. But dare I imagine this concept being nudged into the realm of applicability? In today's world of nanotechnology, quantum mechanics or other related fields (about all of which I know next to nothing), perhaps the notion of a smallest number might find a natural home among those disciplines or elsewhere.

On the other hand, this slim volume may only provide a good chuckle when all is said and done. That's a possibility too. But in any case, as I am a creative writer at heart, I hope to present my idea in an entertaining way rather than offer up cold numbers on a page as I had done years ago, clueless and fresh out of college, when seeking opinions on this matter. A bit more on that later.

First, a brief background on why I began thinking about the existence of the smallest number. It was in late 1986 or early 1987, as best I can remember, that I was reflecting upon the derivation of pi (π). I mean, what math lover doesn't do that now and then in his free time, right? I was working as a substitute teacher in those years, most often covering for math and English courses, so my mind must have stayed in mathematics mode during my off hours, thus prompting me to ponder the peculiarities of pi.

I especially enjoyed math classes while in high school, completing courses in algebra, geometry, trigonometry, advanced algebra and calculus. In college I picked up a math minor just because of my love of numbers, though at this point in my life I've forgotten a good chunk of what I had learned. Yet I still delight in the logic and certainty of numbers and equations as I did in the past, though there were plenty of moments back then which fostered excitement and curiosity as well whenever a new topic was introduced.

One instance occurred during a tenth-grade geometry class in the late 1970s. After having learned in a previous lesson about bisecting a given angle with a straightedge and compass, our teacher had posed a question in the process of showing us how to solve a problem on a related topic, though I can't recall the specific example. But in response, a student in the room jokingly called out to the teacher that he could *trisect* the angle to help solve the problem.

Though amused by the answer, our teacher pointed out that there was no way to trisect an arbitrary angle using a straightedge and compass, the only two instruments allowed under the rules for geometric constructions. In fact, and this was what really caught my attention, he said that there was a proof to show that it couldn't be done.

I remember thinking to myself then and there that I was going to trisect an angle even though I was informed only seconds earlier that there was a proof it couldn't be done. So, over the ensuing days and months (and even off and on when the mood struck me during my college years), I had gone through reams of scrap paper and sharpened many pencils in various and valiant, though ultimately futile, attempts to trisect an arbitrary angle. But through it all, while probably even knowing in the back of my mind that I wouldn't succeed in this task, it was still a lot of fun to try while exercising my math muscles at the same time.

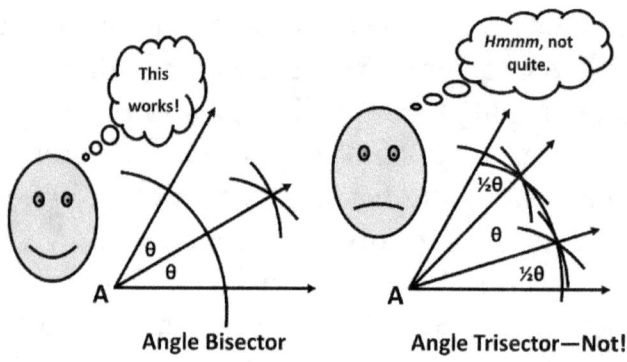

Angle Bisector   **Angle Trisector—Not!**

Similarly, it was this strong sense of interest in all things mathematics which had me pondering the derivation of pi that one day. I don't know why the topic had popped into my head then. Perhaps I had referred to pi while substitute teaching. But I was thinking about the process of approximating pi using an ever-expanding regular polygon inscribed inside a circle. (Look up Archimedes' method to calculate pi for his precise process.)

To approximate pi this way, calculate the perimeter of a regular inscribed polygon (with help from the Pythagorean Theorem to first calculate an individual side) and divide it by the circle's diameter. But to get better approximations of pi, you must inscribe polygons with an increasing number of sides. I realized that by doing this, the most accurate approximation would be obtained when there was the smallest angle possible between two adjacent radii.

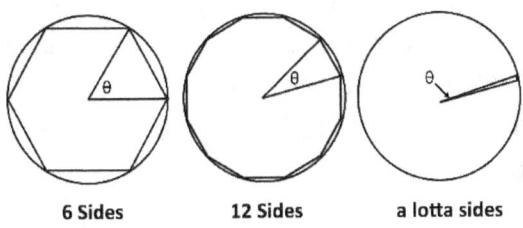

6 Sides   **12 Sides**   **a lotta sides**

Regular Polygons Inscribed in a Circle

(with one side of each polygon connected by two adjacent radii)

Right about now you may be starting to think–*Hey, what's any of this got to do with the topic of your book? Imagining the smallest possible number, remember? Get on with it already!* Trust me, we're almost there.

I recall visualizing an angle θ (theta) between two radii getting smaller and smaller. I imagined θ getting as small as possible, knowing such an angle was numerically *impossible*. Suddenly, a mind-bending idea had wrapped itself around my brain and wouldn't let go. My interest in approximating π vanished in a heartbeat. And that mind-bending idea was this–*What if there existed a smallest angle?* That is, an angle so small that the only smaller angle in existence would have 0°, or in other words, be a straight line.

To the mathematically wired sections of my brain, that was potent stuff. And for the next day or two, I made every attempt to expand upon the idea of the existence of a smallest angle which I named θ, sketching theta-based vertical angles, F patterns and Z patterns, and plugging θ angles into various trigonometric formulas to see if anything would blossom.

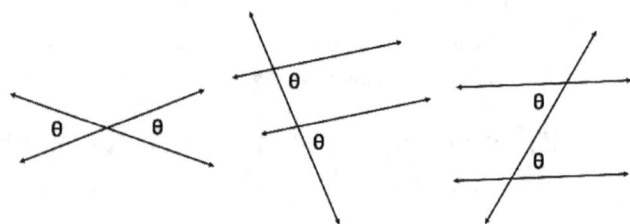

Angling for some inspiration.

But, alas, examining the concept of a smallest angle from every other angle produced nothing of substance, though the cogs in my mind kept turning.

You may see where this is ultimately leading. And though I'm sometimes slow on the uptake, in the next day or so, I eventually saw it, too. While still pondering the idea of the smallest angle, imagining the number of degrees getting as close to zero as possible, imagining an endless string of zeros to the right of a decimal point followed by a one and a degree symbol (°), yet still having to add an infinite quantity of more zeros within that string and *still* not get there, it suddenly struck me that I was looking at this concept all wrong. I had to stop thinking *geometrically*.

Once I dismissed the idea of a smallest *angle* and instead examined the notion of a smallest *number*, everything quickly began to fall into place as I took my new idea and ran with it–right into the proverbial brick wall. *Smallest number? Largest number? No more infinity, but not quite?* Sadly, my allotment of gray matter wasn't sufficient, and still isn't, to help me scale *that* wall. (Plus, my brain begins to hurt when I think about this topic for too long at a stretch.) So, I decided that my best course of action after all these years was to write this little book and set my ideas adrift for others to find, come what may.

Now, let me explain what I have so far.

CHAPTER ONE

Alpha Basics

α

Imagine that there existed the smallest number possible. Let's call this new number α (alpha).

So, precisely where on the real number line (or perhaps on some other number line) would α call home?

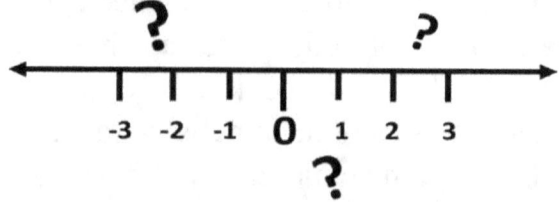

Let the search begin.

First, pretend that you have permission to use the world's most powerful electron microscope. (Hey, you might know people.) When the device is up and running, you focus the beam on a number line (we'll pretend that's also doable) and center it upon zero. So far, so good.

When you look at the generated image on a computer screen, you see the integers sitting on either side of zero. As you gradually focus in, more and more fractions, decimals and roots are revealed whose values grow smaller the closer you zero in on, well, *zero*.

Yes, it's certainly getting crowded on this number line where all the digits hang out. Hey, there's √2 and π talking to each other, both going on and on and on... And you'll notice that .00527 is particularly cranky today, being perpetually wedged in between .00526 and .00528 and letting them know it.

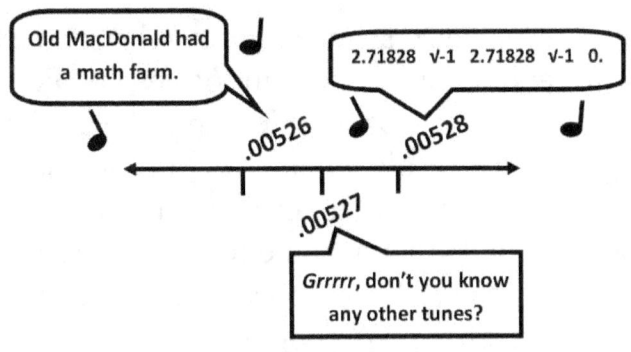

And this numerical sardine tin only gets more crowded the higher the magnification increases. But you persist, diving deeper into the sea of numbers as if inside a tiny submarine. And then, when nearly all hope of ever reaching zero is gone and you're ready to turn back, you finally see it! There, glistening like a tiny lustrous pearl, is alpha, the smallest number possible, sitting shoulder-to-shoulder with zero itself.

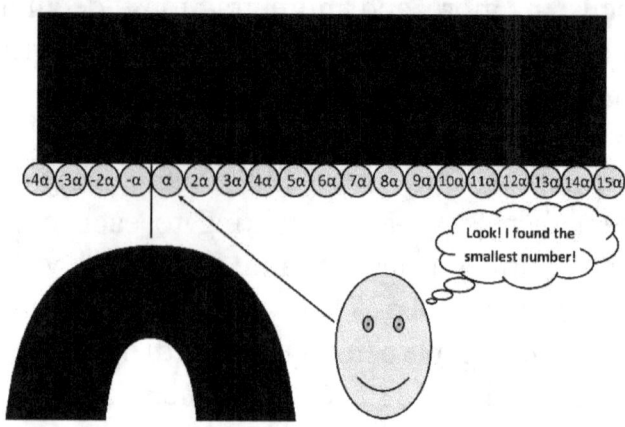

Alpha also sits next to 2α, who sits next to 3α, and so on to the right of zero as far as the eye can see, together looking like a string of pearls. To zero's left are -α, -2α, -3α and on and on. But as your amazement subsides, you start to wonder about something. Exactly how many alphas are lined up from zero to one? A billion? A trillion? A googolplex?

You start to count. And count. And *count*! And you quickly learn that you're getting nowhere fast. It appears that there sure are a lot of alphas lined up here, an infinite quantity perhaps, and yet maybe not.

You soon devise a new plan of action and redirect the electron beam, focusing it on the numeral one on the number line. To your delight and astonishment, you can now see exactly how many alphas are contained within one unit of the number line, namely, K. There are K alphas lined up between zero and one.

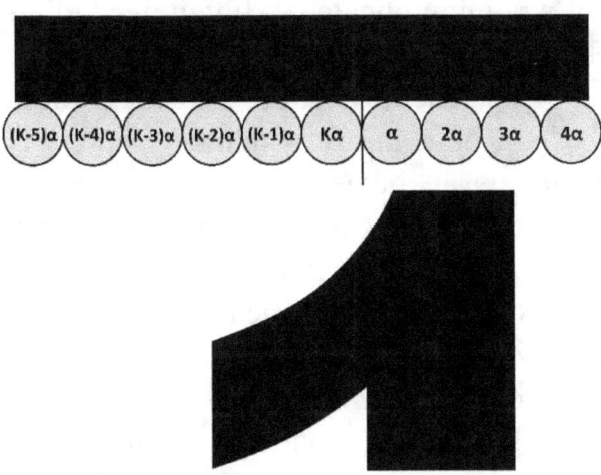

Similarly, you can see that there are K alphas between the numbers one and two, and two and three, (and between their negative counterparts, too) and on and on until–*where*? But before you answer that, you realize that if the number of alphas in any one-unit measure on the number line is always K, then $K\alpha = 1$.

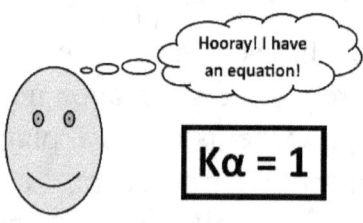

Then another thought pops into your head. If α is the smallest number in existence by its very definition, then is K the largest number? That proposition requires proving, but it's not complicated. All you need to do is look at $K\alpha = 1$ as an inverse variation.

The general equation $xy = c$, where x and y are both nonzero numbers and c is a constant, is an inverse variation. When x is increased (or decreased) by a certain factor, then y must be proportionally decreased (or increased) by that same factor to still obtain c when the new terms are multiplied together. For example, $4 \times 12 = 48$. If 4 is multiplied by **2**, then 12 must be divided by **2** to keep the equation equal. You'll now have $(4 \times \mathbf{2}) \times (12/\mathbf{2}) = 48$, or $8 \times 6 = 48$.

Looking again at $K\alpha = 1$ as an inverse variation, if K is increased by a factor of **2** (or by 3 or 4 or whatever), then α must also be decreased by that same factor of **2** (or by 3 or

4 or whatever) to still obtain 1 when the two new numbers are multiplied together.

But wait! Alpha, by its definition, is already the smallest number possible, so it cannot be decreased by *any* factor. Therefore, K cannot be increased by any factor and will always stay the largest number. Any number smaller than α and larger than K cannot exist in this new number system.

If that is the case, then what about the existence of a larger *quantity* than K, for example, K + 1, sitting on the number line? Such a number can't exist either, and this can be explained by examining the quantity's reciprocal, namely, $1/(K + 1)$.

To do this, let's first look at the fraction $1/K$ by itself. If we increase the denominator while leaving the numerator fixed at 1, the resulting fraction will be smaller, much like

1/9 is smaller than 1/8. With that in mind, we'll go back to our equation $K\alpha = 1$.

If we divide each side by K, we will get $\alpha = 1/K$. Now, if we add 1 to the denominator on the right side, we'll get the reciprocal we considered earlier, namely, $1/(K + 1)$. Since we increased the denominator and left the numerator alone, then the resulting quantity must be smaller than α, the original quantity that $1/K$ is equal to. But that contradicts the definition that alpha is the smallest number possible. Therefore, K + 1 (or K plus any value greater than zero) cannot exist because its reciprocal would be smaller than alpha. This rule would apply to the negative side of zero, too.

So, where does that lead us? Certainly not to the real number line since that extends to infinity in either direction. The number line containing alpha and K has no arrows, but rather endpoints, and could look like this.

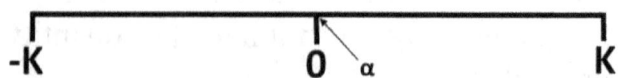

Even though this number line doesn't go to infinity, since K and -K are the largest numbers possible on either end of the system, there is still an element of infinity within if you try to count from zero to K (or even from zero to one). Counting by alphas would be a never-ending task, as would counting backward by alphas from K to zero (or from K to some other number). So, even though there are definite bookends to this system, a component of infinity still exists within it.

To represent other numbers on the line more accurately, the use of a broken number line might offer a better way to go.

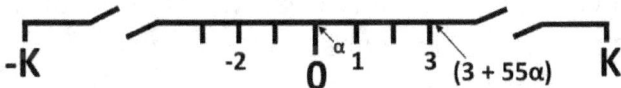

Yet you're still left with two questions. The first is–*What should this new number system be named?* As you're dealing with incredibly small and large numbers, you'll need a word that incorporates both qualities. Should we combine the two words, small and large? *Smarge? Largall?*

It's tricky to find a word describing two words that are opposites, namely, smallest and largest. But perhaps one word exists that illustrates their extreme *degree* of size. While pondering this, I momentarily left the sciences and hopped over to the arts to discover the perfect word–*superlative*.

And we don't want definitions of the word meaning *outstanding* or *exceptional*, but will rather employ the word as used in grammar, that is, comparing three forms of adjectives and adverbs to their varying degrees. (I'll bet that sounds exciting, *huh*?)

Let's pick any old adjective (a word that describes a noun) out of a hat to better explain this. In fact, we'll choose the word *old*. *He is old.* In this sentence, *old* is the adjective in its basic form, called its positive form. *He is older than Bill.* The word *older* is now in the comparative form as it indicates a comparison between two things. *He is the oldest person in the crowd.* The word *oldest* now is presented in the superlative form, a comparison of three or more things.

Yeah, I get it!

POSITIVE	COMPARATIVE	SUPERLATIVE
Small	Smaller	Smallest
Large	Larger	Largest

The word *superlative*, when used this way, perfectly describes a number system containing the smallest and largest numbers. But our second question still looms, and that simply is this–*What's next*?

CHAPTER TWO

Throwing Superlative Numbers
Against the Wall and Hoping
Something Sticks

?

Yes, it's a very good second question. What *is* next? In the previous chapter we had poured the concrete foundation for what I've named the superlative number system. But can we build something upon it?

Pondering the existence of the smallest and largest numbers might ultimately prove to be just a fun mental exercise, an intellectual diversion good for a few mathematical laughs. And that's fine. But if this concept can be expanded upon somehow, what's the best way to proceed? What, if anything, can we do with superlative numbers?

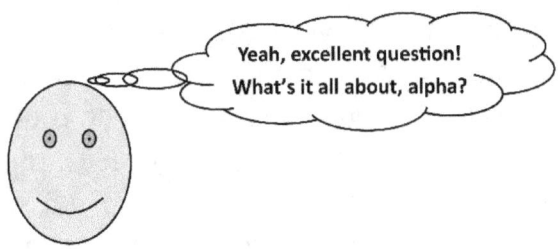

Yeah, excellent question! What's it all about, alpha?

LET'S NOT LIMIT OURSELVES

I thought about this over thirty years ago and decided to examine these numbers as they might relate to limits. Alpha is the smallest superlative number, sitting shoulder-to-shoulder with zero. Likewise, K is the largest number, the nearest thing to the concept of infinity in this new system. So, I searched though my calculus textbook for a few limits involving infinity and wondered how they might change when introducing α and K into the mix.

THREE LIMITS AS $n \rightarrow \infty$	1) $\lim\limits_{n \to \infty} x^{1/n} \rightarrow 1$	when $x > 0$
	2) $\lim\limits_{n \to \infty} x^n \rightarrow 0$	when $|x| < 1$
	3) $\lim\limits_{n \to \infty} x^n \rightarrow \infty$	when $|x| > 1$

In the first limit above, as n approaches infinity, the exponent $1/n$ will approach zero. But in the superlative number system, α is the closest number to zero, so once $1/n$ reaches α, it can no longer *approach* zero since it is now as close to zero as it can get. So, let's replace exponent $1/n$ with α. We can also get rid of the $n \rightarrow \infty$ term now since our new exponent is no longer *approaching*, but has arrived.

Similarly, in the second and third limits above, the exponents n will approach infinity. But in the superlative number system, K is the closest equivalent to the concept of infinity, so once n reaches K, it can no longer *approach* infinity since it is now as far from zero as possible. So, let's replace exponents n with K. We can also get rid of the $n \rightarrow \infty$ terms now since our new exponents are no longer *approaching*, but have arrived. We're not done, but so far, we have this.

	1) $\lim x^{\alpha} \rightarrow 1$	when $x > 0$
THREE LIMITS		
UNDER	2) $\lim x^{K} \rightarrow 0$	when $|x| < 1$
RENOVATION		
	3) $\lim x^{K} \rightarrow \infty$	when $|x| > 1$

Now we'll examine the x values. In the first limit, since x > 0, we can test both α and K as x values since each are greater than zero. This will create $\lim \alpha^\alpha \to 1$ and $\lim K^\alpha \to 1$. Since the bases and exponents in each limit can't get any smaller or larger now, then α^α and K^α will approach one as closely as possible from the left and right, respectively. The closest α^α can get to 1 is one alpha unit away to the left, or $1 - \alpha$. The closest K^α can get to 1 is one alpha unit away to the right, or $1 + \alpha$.

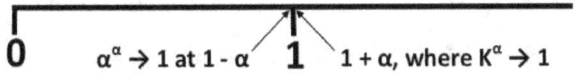

0    $\alpha^\alpha \to 1$ at $1 - \alpha$ **1** $1 + \alpha$, where $K^\alpha \to 1$

α^α and K^α approach 1 at closest points from the left and right, respectively, at points 1 - α and 1 + α

And since α^α and K^α are as close to one as they can ever get, we'll replace the arrows with equal signs and eliminate the lim notation to get $\alpha^\alpha = 1 - \alpha$ and $K^\alpha = 1 + \alpha$.

In the second limit, since $|x| < 1$, we can test α as an x value, as α is less than one, to get $\lim \alpha^K \to 0$. In the third limit, since $|x| > 1$, we'll test K as an x value, as K is greater than one, to get $\lim K^K \to \infty$. Let's also replace 0 and ∞ with α and K, respectively, since α is the closest number to zero that α^K can approach and K is the largest number, the closest thing to infinity in the superlative number system that K^K can approach. What we have left is $\lim \alpha^K \rightarrow \alpha$ and $\lim K^K \rightarrow K$.

And as in the first limit, since the bases and exponents in the second and third limits cannot get any smaller or larger, we'll replace each arrow with an equal sign and eliminate the lim notations to get $\alpha^K = \alpha$ and $K^K = K$. Here are our newly created equations.

NEW EQUATIONS CREATED FROM OLD LIMITS

$$\alpha^\alpha = 1 - \alpha \quad K^\alpha = 1 + \alpha$$

$$\alpha^K = \alpha \quad K^K = K$$

I've tried to take a logical approach when creating these equations, though they shouldn't be considered as proven true in a mathematical sense. I'm more interested in the concept right now. (And I'm not sure how I'd prove them true, or not, anyway.) But let's go with the results and see where they lead us. After all, this chapter is titled *Throwing Superlative Numbers Against the Wall and Hoping Something Sticks*, so I'm doing just that.

SOME SECOND THOUGHTS

After thinking about the two equations $\alpha^K = \alpha$ and $K^K = K$, I reexamined some of the conclusions I had made thirty-plus years ago. Let's first look at $\alpha^K = \alpha$ more closely.

Any number greater than zero and less than one raised to a power greater than one will result in a smaller number. For instance, $.5^4 = .0625$. The resulting $.0625$ is less than $.5$, the original value. But since α is the smallest number possible, raising α to *any* number greater than one can't result in a smaller number since such a number doesn't exist. So, should α^K be considered undefined instead of being equal to α? Or, is alpha raised to *any* power greater than one *always* equal to alpha?

When I first considered this matter in the mid-1980s, I regarded α^2 as undefined, much like dividing a number by zero is labeled as undefined. But if $\alpha^K = \alpha$ is to hold true, then so must other similar expressions.

ALPHA RAISED TO A POWER GREATER THAN ONE IS ALWAYS EQUAL TO ALPHA
$\alpha^2 = \alpha \quad \alpha^{30} = \alpha \quad \alpha^{1000} = \alpha \quad \alpha^K = \alpha$

Now let's focus on $K^K = K$. Raising a number greater than one to a power greater than one will always result in a larger number than the original. For instance, $8^3 = 512$. The resulting 512 is greater than 8, the original value. But since K is the largest number possible, raising K to *any* power greater than one can't result in a larger number since such a number doesn't exist.

So, as in the previous example, should K^K be considered as undefined instead of being equal to K? Or, is K raised to *any* power greater than one *always* equal to K? Again, we'll go with the latter argument. If $K^K = K$ is to hold true, then so must other similar expressions.

K RAISED TO A POWER GREATER THAN ONE IS ALWAYS EQUAL TO K

$K^2 = K$ $K^{30} = K$ $K^{1000} = K$ $K^K = K$

EXAMINING Kα AS AN EXPONENT

While throwing superlatives against the wall years ago, I noticed something else. Though Kα = 1, if Kα is used as an exponent on one side of an equation, replacing it with 1 renders the equation unbalanced. Here are examples using two of our four newly created equations again listed below.

NEW EQUATIONS CREATED FROM OLD LIMITS

$$\alpha^\alpha = 1 - \alpha \quad K^\alpha = 1 + \alpha$$

$$\alpha^K = \alpha \qquad K^K = K$$

We'll first look at $\alpha^K = \alpha$, raising each side of the equation to the power of α.

We'll examine equation	$\alpha^K = \alpha$
Raise each side to the power of α	$(\alpha^K)^\alpha = \alpha^\alpha$
Multiply exponents on left side	$\alpha^{K\alpha} = \alpha^\alpha$
Let exponent $K\alpha = 1$ on left side	$\alpha^1 = \alpha^\alpha$
Set $\alpha^1 = \alpha$ on left side	$\alpha = \alpha^\alpha$
Set $\alpha^\alpha = 1 - \alpha$ on right side	$\alpha = 1 - \alpha$ is not true

By substituting 1 for the exponent $K\alpha$, we've created an untrue statement. Now, let's examine $K^K = K$, again raising each side to the power of α to see if we get a similar result.

We'll examine equation	$K^K = K$
Raise each side to the power of α	$(K^K)^\alpha = K^\alpha$
Multiply exponents on left side	$K^{K\alpha} = K^\alpha$
Let exponent $K\alpha = 1$ on left side	$K^1 = K^\alpha$
Set $K^1 = K$ on left side	$K = K^\alpha$
Set $K^\alpha = 1 + \alpha$ on right side	$K = 1 + \alpha$ is not true

Wow! That was a bit dizzying! What's our next step?

Our next step is to reexamine the same two equations without substituting 1 for the exponent $K\alpha$ and see what happens. We'll look at $\alpha^K = \alpha$ first.

We'll examine equation again	$\alpha^K = \alpha$
Raise each side to the power of α	$(\alpha^K)^\alpha = \alpha^\alpha$
Since $\alpha^K = \alpha$, replace α^K with α on left side	$(\alpha)^\alpha = \alpha^\alpha$
Set $(\alpha)^\alpha = \alpha^\alpha$ on left side	$\alpha^\alpha = \alpha^\alpha$
Set $\alpha^\alpha = 1 - \alpha$ on both sides	$1 - \alpha = 1 - \alpha$ which is true

So, the equation $\alpha^K = \alpha$, after raising each side to the power of α, can be solved when exponents K and α are left in their original forms. Let's now reexamine $K^K = K$ in the same manner, again raising each side to the power of α without substituting 1 for exponent $K\alpha$. Feel free to skip over the next chart if one example was enough.

We'll examine equation again	$K^K = K$
Raise each side to the power of α	$(K^K)^\alpha = K^\alpha$
Since $K^K = K$, replace K^K with K on left side	$(K)^\alpha = K^\alpha$
Set $(K)^\alpha = K^\alpha$ on left side	$K^\alpha = K^\alpha$
Set $K^\alpha = 1 + \alpha$ on both sides	$1 + \alpha = 1 + \alpha$ which is true

Success! This theory holds water for the moment. But let's look at both equations a final time, rearranging exponents on the left to make sure if both equations can still be solved without the substitution of 1 for $K\alpha$.

We'll examine equation a third time	$\alpha^K = \alpha$
Raise each side to the power of α	$(\alpha^K)^\alpha = \alpha^\alpha$
Rearrange exponents on left side of equation	$(\alpha^\alpha)^K = \alpha^\alpha$
Since $\alpha^\alpha = 1 - \alpha$, replace α^α with $1 - \alpha$ on both sides	$(1 - \alpha)^K = 1 - \alpha$ is not true

We're back to square one again as the equation is not solved. Since $1 - \alpha$ is less than one, raising it to the power of K will produce an incredibly small number not equal to $1 - \alpha$. We can look at our second example, $K^K = K$, raising each side of the equation to the power of α and rearranging the exponents. Will we get a similar result as in the above example? Again, feel free to skip over the next chart if the previous example did the trick.

We'll examine equation a third time	$K^K = K$
Raise each side to the power of α	$(K^K)^\alpha = K^\alpha$
Rearrange exponents on left side of equation	$(K^\alpha)^K = K^\alpha$

Since $K^\alpha = 1 + \alpha$, replace K^α with $1 + \alpha$ on both sides	$(1 + \alpha)^K = 1 + \alpha$ is not true

Just like in the previous example, the equation cannot be solved when rearranging the exponents even though they were left in their original forms. Since $1 + \alpha$ is greater than one, raising that to the power of K will produce an incredibly large number clearly not equal to $1 + \alpha$.

But being persistent, I slogged on and posed this question to myself. Will raising an equation to the power of $1/K$ or $1/\alpha$ produce similar results if one side of the equation is to the power of K or α, respectively? Since the new exponent would be either K/K or α/α, both equaling one, will an untrue statement result if the exponent is set equal to one before solving? And if not set equal to one, will things break down anyway as in the last two examples when trying to solve?

I won't keep you in suspense for much longer and promise not to make your eyes glaze over by presenting several more charts of mind-numbing equations. We get a mixed bag of results, just as before. When terms such as $K\alpha$, K/K or α/α are used as exponents, they cannot first be reduced to 1 to help solve these problems. Yet when working with $K\alpha$,

K/K and α/α while in their original forms, we'll sometimes be able to solve the problems correctly and sometimes not.

SOME FINAL QUESTIONS

When I started writing this book and dug a little deeper into superlative numbers, an intriguing question occurred to me. Since the alpha numbers run shoulder-to-shoulder from -K to K on the superlative number line, will every existing number have a position on this line, or will some cease to exist?

-K   0   K

There is seemingly a place for every number–for 1, for 547 + 98α, for -3000α, for 28.719 and even for 783.036 + 102α. So, and this is the intriguing question, can irrational numbers exist in this system? For example, can $\sqrt{2}$ and π, which both go on forever, or repeating decimals such as 2.33333..., exist in the superlative number system?

Irrational numbers have no ending digit, but every point between -K and K on the superlative number line is either a whole number (e.g. 57,602), a whole number with a finite decimal attached (e.g. 982.6412) or a whole multiple of alpha with or without the previously noted whole number attachments (e.g. 37α, $76 + 4\alpha$, $11.35 + 2\alpha$). And since the latter kind takes up all the space between the first two types, there is no room between any adjacent alpha pair for another number to slip through and wind on infinitely.

So, must irrational numbers on the real number line be truncated or transformed to be allowed to hang out in the superlative system? And if that isn't possible, must they otherwise disappear?

I don't have answers to these questions, but what has to happen to irrational numbers for them to exist in this realm? Must they now all have a last digit, whether being expressed with or without an alpha component? Will pi finally reach an endpoint when, or if, it can be derived in this new system and/or in terms of alpha? Or will values such as pi and $\sqrt{2}$ just cease to exist in their current forms in the superlative system? Most importantly, can the answers to these and other such thrilling questions *ever* be found?

A FINAL THOUGHT

The key points in this brief text are that alpha, by its very definition, is the smallest existing number, and that K, therefore, is the largest, all leading to the equation $K\alpha = 1$. These new numbers, which comprise the superlative number system, run from -K to K, with a string of alpha components running shoulder-to-shoulder from one end to the other, seemingly infinite, yet with endpoints on either side of the number line at -K and K.

And after reconfiguring a few limits involving infinity in terms of alpha and K, I've proposed four more equations, namely, $\alpha^\alpha = 1 - \alpha$, $K^\alpha = 1 + \alpha$, $\alpha^K = \alpha$, and $K^K = K$.

The few other ideas mentioned in this book are based on the above as a foundation.

Admittedly, the other ideas are few and just pulled randomly out of the air as I am at a loss about how to rigorously advance my basic idea, if it can even be done. Are there other ways to graph superlatives? Apply them to other limits? Other mathematical formulas? Manipulate them geometrically? Explore them in different ways I cannot even imagine or understand should others find a better approach? Or should some or all of my basic assumptions be reconfigured altogether?

If others show any interest in further pursuing such a line of thought, I say go for it. I'd be curious to read someday where others' ideas have taken them. In the meantime, I'm just happy to get this particular *What if...?* of mine, this nugget of an idea, finally set down in a (hopefully understandable) written form. And that, at long last, is good enough for me.

__CONCLUSION__

Then and Now

    This is still a small book about a very small idea. (And a very large idea, too.) And maybe that's the most definitive statement I can make about the concept of alpha, K and the superlative number system–that it is only an *idea*. And admittedly, an idea not fully formed. And definitely one needing a dash of higher brainpower to shape it into something more substantial.

    But if too many incomprehensible and contradictory elements exist within such a system, then maybe this is, at best, just a fleeting idea, a soap bubble of a notion to while away a bit of your time. But whatever the case may be, we've nearly made it to the end of this mathematical journey. Well, at least *I* have anyway.

After thinking about all of this from time to time over the last thirty-plus years, my mental gears have ground to a halt trying to make sense of, or advancing, such a concept. Putting boundaries, bookends or limits (pun intended) on infinity may be a mathematical undertaking that isn't possible. Might as well try to count raindrops during a thunderstorm.

Still, I wanted to get this idea written down in some form to purge it from my cluttered mind so I can get on with other writing projects. In doing so, I've included a few snippets of my original notes from years ago when I presented this idea in a more formal way, a more *mathematical* way, which I thought was the appropriate way to go to garner other opinions.

I recall searching for titles of several math journals, trying to find one or two whose content I thought might relate to my idea. I had no clue about the proper presentation of scholarly work and sent off letters and my typed notes to a few editors and others in the mathematics field, hoping for a comment or two on my concept.

Some kindly replied, though not fully grasping my basic idea, but I'll attribute that to me not being clear in my presentation (unlike *this* clearer presentation, or at least as clear as my brain can make it). Below is an excerpt written in the mid-1980s where I introduced the concept of a smallest number.

1) Let ε be the smallest real number. In other words, no value x exists for $0 < x < \varepsilon$. The statement that ε is the smallest real number should be taken as a <u>definition</u> of ε. It is the basis for the entire number system. The quantity ε cannot be reduced, halved or broken down in any manner because there is nothing smaller than ε except zero. Any expression alluding to this, such as $\varepsilon/2$ or ε^2 should be considered <u>undefined.</u>

2) Let k be the number of times ε is divisible into 1. That is, $\frac{1}{\varepsilon} = k$. Therefore $\varepsilon k = 1$.

The first thing you'll notice is that I had used ε (epsilon) instead of α to symbolize the smallest number. I also utilized a lowercase k instead of capital K as the largest number. It wasn't until late last year, after deciding to write this up in its current form, that it finally dawned on me to research online to see if anyone had explored this topic along the same lines that I had, but I came up empty.

But while examining a list of Greek letters used as mathematical symbols, I found that epsilon, coincidentally, was defined as an arbitrarily small number close to zero, an infinitesimal quantity. Unlike my definition of alpha though, epsilon is not a fixed quantity on a number line. To avoid confusion, I had considered renaming my invented number ε' (epsilon prime) before eventually settling upon alpha. And since K is the largest superlative number, I splurged a few extra bucks and went with the uppercase version, which seemed a more appropriate choice.

Another change from my original notes which was addressed in the last chapter is that I no longer consider the expression ε^2 (rather now α^2, or α raised to any power greater than one) as undefined.

ALPHA RAISED TO A POWER GREATER THAN ONE IS ALWAYS EQUAL TO ALPHA

$$\alpha^2 = \alpha \qquad \alpha^{30} = \alpha \qquad \alpha^{1000} = \alpha \qquad \alpha^{K} = \alpha$$

The most drastic change, however, is one that I made while preparing this text. I reconsidered two of my four invented equations, setting $\alpha^\alpha = 1 - \alpha$ and $K^\alpha = 1 + \alpha$. Originally, and I now consider mistakenly, I had set both α^α and K^α equal to one (or rather ε^ε and k^ε both equal to one as first written in the notes below). But as both terms were *approaching* one and not equal to one, I adjusted accordingly.

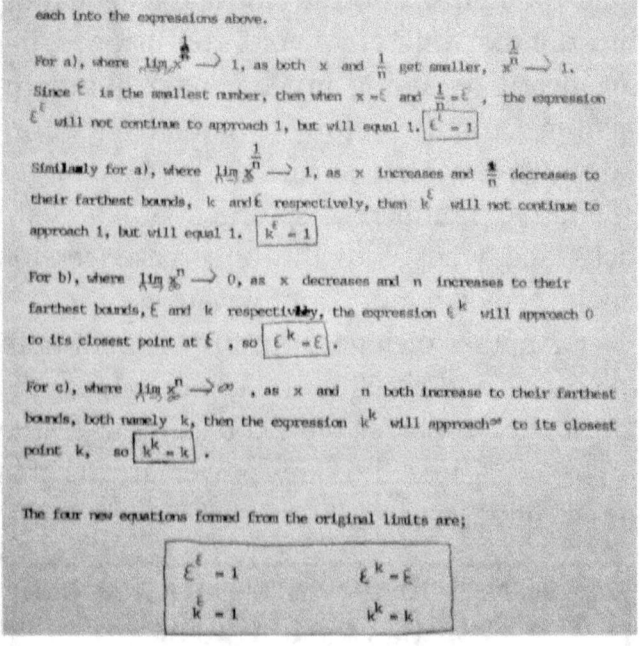

These next excerpts, and the previous ones, also highlight another interesting feature from the late 1980s and earlier–that charming, finger-staining copying device called carbon paper.

These images of my notes are photos of the *copies* I had made for myself using carbon paper. For any younger readers who might not have heard of the delight that is carbon paper, it is a flimsy sheet of plastic-like paper, slick on one side and covered in a dried ink blend on the other. To copy what you were about to type, you'd sandwich a sheet of carbon paper between two sheets of regular paper and then insert the trio into the typewriter ink-side down. As a typewriter key made an original impression on the top sheet of paper, it also imprinted a copy on the second sheet by transferring a bit of the ink from the carbon paper via the force of the pounding key.

In the images below, the tips of my approaching arrows and the absolute value lines are among two misaligned items. As the typewriter keys didn't contain all the symbols I required, I'd leave spaces where needed and draw in the missing images later. But the two sheets of regular paper, with the carbon paper between them, rarely lined up perfectly when removed from the typewriter, so my carbon copy drawings were usually a bit off, though some pages look better than others.

4) Recall the limits

a) $\lim_{n \to \infty} x^{\frac{1}{n}} \to 1$ when $x > 0$.

b) $\lim_{n \to \infty} x^n \to 0$ when $x < 1$.

c) $\lim_{n \to \infty} x^n \to \infty$ when $x > 1$.

Since ϵ and k are the bounds between 0 and ∞ respectively, substitute

6) When used as an exponent, the expression ϵk must be left in the form ϵk. For example, $\epsilon^{\epsilon k} = \epsilon^{\epsilon}$, cannot be rewritten as $\epsilon^1 = \epsilon^{\epsilon}$ since $\epsilon^1 \neq \epsilon^{\epsilon}$.

7) Cancelling fractions in exponents involving ϵ and k cannot occur. Note that $\epsilon^k = \epsilon$ from Section 4. Now raise each side to $\frac{1}{k}$ to get $(\epsilon^k)^{\frac{1}{k}} = (\epsilon)^{\frac{1}{k}}$. If k and $\frac{1}{k}$ were cancelled, you'd have $\epsilon^1 = \epsilon^{\frac{1}{k}}$ which is $\epsilon^1 = \epsilon^{\epsilon}$, or $\epsilon^1 = 1$ which isn't true. Instead let $\frac{1}{k} = \epsilon$ to get $\epsilon^{k\epsilon} = \epsilon^{\epsilon}$, or $1 = 1$.

But making manuscript copies that way was par for the course back then. The process seems more inconvenient in retrospect than it probably actually was. But that's how it is with most things when comparing then to now. Today it's easy to type, edit and print clean copies compared to other eras. And nowadays, who even needs to print paper copies with texting, email and the like? Still, compared to the trouble people had to endure in the time of Dickens or before Gutenberg, carbon paper seems downright high-tech.

I'll admit though that making copies of some of the novels I wrote years ago of a few hundred pages each (while reusing a sheet of carbon paper a dozen or so times until the ink was worn bare) *was* a bit of a challenge back then—especially after multiple typos had to be corrected. For that we had white correcting tape, among other items. But that's a whole other tangent.

Let's circle back to mathematics and wrap things up with this quote attributed to Albert Einstein. *If at first the idea is not absurd, then there is no hope for it.* And while

I'm not entirely certain if I'd label the notion of a smallest and largest number as absurd, it is a bit out there on the edge, to say the least. So, I guess that's a good sign.

THE END

AUTHOR'S NOTE

Thanks very much for reading this book, and even if all the contents didn't make complete sense, I hope I at least successfully conveyed my main points.

This publication marks my first non-fiction title. All my other works are novels of various lengths and genres. But if you like a bit of mathematics mixed with your fiction, you may enjoy reading two of my other books.

The first one, *The Question Lies on Varick Street*, is the second entry in my 22-Minute Novel series for adults and older teens. It uses the Fibonacci sequence as part of its plot. The other title, *Search for the Silver Swamp Monster*, Book #1 in the Griffin Ghostley Adventure Series for readers ages 10 to 13, also contains an element of math crucial to the story. But I'll reveal no more here.

Thanks again. And may all your problems be alpha size and all your successes feel as big as K. Take care!

Thomas J. Prestopnik
May 31, 2018

BOOKS BY
THOMAS J. PRESTOPNIK

NON-FICTION

Zero's Next-Door Neighbor
Imagining the Existence of the Smallest Number,
the Largest Number, and the Superlative Number System

FICTION

The Griffin Ghostley Adventure Series
for readers ages 10 to 13
Search for the Silver Swamp Monster #1
Prisoner of the Giant Boona Bird #2
The Snow Beast of Finnian Forest #3

Missing Tweets no. 4
The Question Lies on Varick Street

for adults & older teens

Nicholas Raven and the Wizards' Web
an epic fantasy in three volumes
for adults & older teens

A Christmas Castle
a novella for adults & older teens

The Endora Trilogy
a fantasy-adventure series for pre-teens & adults

The Timedoor - Book I
The Sword and the Crown - Book II
The Saving Light - Book III

Gabriel's Journey
an adventure novel for pre-teens & adults

Visit Thomas J. Prestopnik's official website
www.TomPresto.com